Ranching –

Ochsner Ranch

George and Ruby Ochsner receive a special award from the American Hereford Association. Ochsner Ranch is a five-generation "Family Ranch." The ranch was started in 1913 by Jacob Ochsner and is now run by his grandson, George, and wife, Ruby, along with sons, Blake and Rodney, and daughter and son-in-law, Dixie and Steve Roth. Dixie and Steve's son, Rustin, represents the fifth generation working full-time on the ranch. The 22,000-acre ranch is located 20 miles north of Torrington, Wyoming, and maintains registered herds of Hereford and Angus cattle, a commercial cow operation and a feed yard.

The Ochsners are actively involved in all aspects of beef production. The sixth generation, the Roth's granddaughters, Tansy and Ember, visit Grammy and Grampy's every chance they get to help them on the ranch.

Cross Diamond Cattle Company "The Ford Family"

Cross Diamond Cattle Company, where our dad works, is the "Ranch Family" we are a part of and where we call home. It is owned by Scott and Kim Ford of Bertrand, Nebraska. They are a family business along with their two daughters, Johanna and Marie. Cross Diamond is a Red Angus seedstock operation located in South Central, Nebraska. Cross Diamond is also made up of several families all working together on two ranches, the registered operation and the commercial operation, dedicated to raising top-quality Red Angus cattle. Cross Diamond branding day, above, is when many friends and neighbors come to help. We are proud to be a part of the Cross Diamond family. It's not generations of family members on the same ranch, but rather a combination of hard-working families all "riding for the Cross Diamond brand," and for the Ford family.

To all the Farmers and Ranchers in America,

We salute all the hard-working farmers and ranchers in America, who work tirelessly producing safe, wholesome and nutritious food for consumers in America and overseas. We are proud 97 percent of U.S. farms and ranches are family owned. We thank the two percent of the nation's population who call themselves farmers and ranchers!

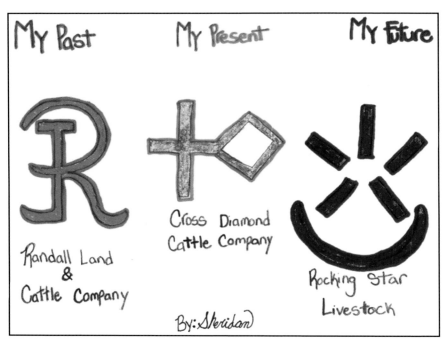

Published by Down Under Publications

All rights reserved. No part of this book either in part or in whole, may by reproduced, transmitted, or utilized in any form, by any means, electronic, photographic, or mechanical, including photocopying, recording, or by any information storage system, without permission from the author, except for brief quotations embodied in literary articles and reviews.

First Edition, December 2016
©Copyright, Rebecca Chaney, 2016
ISBN -978-0-9818468-7-3

Chaneyswalkabout@aol.com
website: www.rebeccalongchaney.com
Blog: chaneytwinsagbooks.blogspot.com
Edited by Rebecca Long Chaney
Photographs by Kelly Hahn Johnson and Rebecca Long Chaney
Guest Photographers - Dixie Roth, Kim Ford & Ryan Overlease
Layout and Design by Kathy Moser Stowers
Lesson Plans for Books #4 through #8 by Laura Keilholtz, Professional Educator.

What a special morning to visit with Mr. George Ochsner and his two sons, son-in-law and grandson. Every morning the three generations gather around the kitchen table at the Ochsner home ranch where George and Ruby Ochsner live in Wyoming to discuss the day's schedule.

Ochsner Ranch started in 1913 by George Ochsner's grandfather, Jacob, which makes it a centennial ranch or 100 years old. Mr. George explains to us the ranch's rich history and shares his love of cattle. We find out the ranch is about 22,000 acres – that's really, really big. Mr. George is 83 years old and still checks 13 pastures of cattle by himself.

We visited the Ochsner Ranch at the beginning of calving season when the momma cows have their calves. The cows that are close to giving birth are moved closer to the calving sheds near the homestead. This bunkhouse by the calving pen is where Ochsner family members sleep every night for weeks keeping a watchful eye on the momma cows in case the animals need assistance.

During unpleasant weather, there are calving sheds for expectant cows. As the weather worsens, cows and calves can be moved inside calving sheds for protection. Ms. Dixie, George Ochsner's daughter, feeds a calf a bottle of milk. Some mother cows are not able to feed their calves so ranchers step in and the calf becomes a bottle calf.

We are so excited the Ochsners have Herefords, because they're our favorite breed. This calf really likes us. The odd looking box is called a warming box. Wyoming can have some extremely cold and windy weather. Warming boxes often help calves get dry after birth and can save their lives.

Ms. Dixie uses a hydro-arm on the back of the truck to pick up large round bales of hay. What an awesome contraption! When we go to the pasture, the hydro-arm lowers the bale, and rolls it out slowly for the cattle to eat. Ms. Dixie loves Herefords as much as we do!

Before the hydro-arm can lower the bale and roll it out, we carefully cut the netwrap off the bale. Hay is grass or legumes grown in summer, dried and baled. It provides feed for cattle during times when there's not enough grass growing in the pastures.

The "Fort" is a large structure where cattle and calves are moved when a dangerous snowstorm is blowing in over the prairie. The Fort has saved many cows and calves during blizzards and snowstorms. We think the Fort would make an awesome place to play.

Solar panels are popping up in the middle of pastures across the west to power cattle water tanks. This solar panel collects the sun's rays and turns them into energy or electricity to pump water into three water tanks to make sure cattle get clean water daily. We learned a lot working with Ms. Dixie on her family's generational ranch and can't wait for our next visit. Now, it's time to get home to our ranch family.

We live on a different kind of "Family Ranch." It is not generations in the making, but rather made up of several families working for the Ford family, owners of Cross Diamond Cattle Company. Ranch life is not all work. When we go check on animals with our dad, we often get to play. Our best times are often spent with Johanna and Marie Ford.

The Cross Diamond ranch family is awesome and we have so many role models. Well, some of the cowboys play tricks on us, like putting plastic snakes in the barn to scare us. Here are owners, Kim and Scott Ford, pictured in the middle. It's an honor to be part of such a hard-working crew.

Calving season is still in full swing at Cross Diamond. While some calves are content to snuggle down in the pasture, others get into mischief and get on the wrong side of the fence. Our dad gently pushes the calf back through the fence to its mother.

Nearly half a million Sandhill Cranes spend March and April in Nebraska during their spring migration preparing for their long journey north to their breeding grounds in Canada and Alaska. We are lucky that some of them stop every year in our pastures and eat alongside the Cross Diamond cattle. Did you know farmers and ranchers provide 75 percent of the nation's wildlife habitat?

Cross Diamond branding day is always one of our biggest days at the ranch. Dad and Andy rope calves for the branding crew. At right, Andy pulls a calf towards the branding crew where it will receive its vaccination and its brand. A brand is a permanent symbol and identifies which rancher owns the animal.

At a smaller branding, we help keep calves back using a livestock sorting stick as Marcus ropes the calves. There are three kinds of brands — a freeze brand, a hot brand and an electric brand. We have worked with all three brands and they are very special to us.

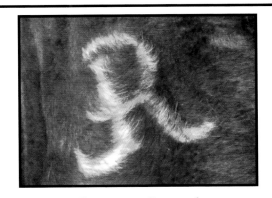

**Freeze Brand
Randall Land & Cattle Company**

The branding iron is cooled to -100 to -300 degrees Fahrenheit and placed on animal for about 60 seconds. This brand is from the Hereford Ranch where we were born in Maryland.

**Hot Brand
Cross Diamond Cattle Company**

Brand is laid in hot fire and put on animal 3 to 4 seconds. This brand is from Cross Diamond Cattle Company where our Dad works today. We are proud to be a part of the Cross Diamond family.

**Electric Brand
Rocking Star Livestock**

An electric brand is plugged in and ready when hot. Stays on animal 3 to 4 seconds. This is our own brand, "Rocking Star," in honor of Little Star, the first Hereford we ever raised.

Once the 437 calves are vaccinated and branded it's time for the big branding lunch. Mr. Scott begins with prayer, thanking God for the many blessings of the day and year. More than 30 friends enjoy a branding lunch and an afternoon of fellowship.

Branding day wraps-up with water balloons and lots of fun. Twenty-one-month-old Coltin is the youngest member of our ranching family and brings us much delight. Another great day with our ranch family.

Fall is a busy time of year. Megan ropes a calf in the pasture that appears to be a little sick while dad doctors it or gives it medicine to feel better. Corey and Marcus sort bulls near the homestead to be videotaped for the upcoming sale in December. Taking the best care of the cattle is the main priority every day on the Cross Diamond ranch.

While Dad gathers bulls on a four-wheeler, Jaxson, our Border Collie, uses his natural herding instinct to work the cattle slowly and calmly. Jaxson is our family dog, but seems to enjoy helping our dad move cattle to the next pasture.

A camera is on this drone or unmanned aerial vehicle (UAV). It just flew over the Cross Diamond Production Sale and took some awesome photos from high in the sky. Drones are a huge management tool in agriculture now and are being used to check crops and livestock in pastures and fields far from the homestead.

The work all year at the ranch culminates in the Cross Diamond Production Sale. A video of each animal pops up on several television screens in the sale barn and an auctioneer calls the bidding while men out front take the bids. Our dad, along with other cowboys, stay with the bulls outside for customers to look at them.

We are blessed to be a part of the Cross Diamond ranching family. From working cattle with Johanna and Marie to just spending time together, we have learned that ranching is more than a commitment to the cattle and the land, but also to the people with whom we share our lives and passion for agriculture.

Dedication and hard work describe ranchers and farmers 365 days a year, but there are also times to celebrate the special occasions. We are so excited to congratulate Corey, one of the cowboys from Cross Diamond, on his marriage to Meredith this year. Our ranching family will continue to be committed to providing healthy beef, as well as being good stewards of the land that has been entrusted to us for generations to come.

Glossary

Acre – Is a measurement of land, a way to measure farm or ranch land.

Agriculture – A fancy word used to describe farming and ranching.

Branding/Brands – Are permanent symbols ranchers give their livestock to identify the animal's ranch beginning.

Calving Season – A term used in the spring when most cows are having their calves.

Calving Sheds – Buildings in western states where cows can deliver their calves during severe weather.

Cross Diamond Production Sale – Scott and Kim Ford hold an annual sale of some of their best cattle.

Generations – If a child is a third generation farmer, that means their parents and their grandparents also farmed.

Habitat – The natural environment or place where wildlife or other organisms live.

Hereford – A popular breed of beef cattle that originated in England.

Hydro-Arm – A mechanical device attached to the back of a truck capable of lifting heavy round bales.

Migration – When animals, usually birds, move to warmer weather to survive the winter months.

Pasture – A fenced-in area where cattle and other livestock graze.

Ranch/rancher – A rancher is a person who raises livestock on a ranch or large portion of land.

Dixie and Steve Roth are excited that granddaughters, Tansy and Ember Assmann, the sixth generation, love the ranch. Some of Tansy and Ember's favorite times are spent with Grammy and Grampy on the Ochsner Ranch. In July, 2016, a devastating wildfire destroyed 11,000 acres of the Ochsner Ranch. Some of the buildings and cattle pictured in this book were lost, but there were also many blessings – houses saved, lives saved. We hope this children's book pays tribute to the Ochsner's incredible perserversance through this tragedy.

A huge thanks to Dixie Roth for contributing a couple of her beautiful photos to the Chaney Twins' book. To order "Dixie Pics," country portraits or greeting cards, email her at sdroth@hughes.net.

10th Anniversary Cross Diamond Production Sale and all the hard work at the cattle ranch all year is celebrated!

December 14, 2015

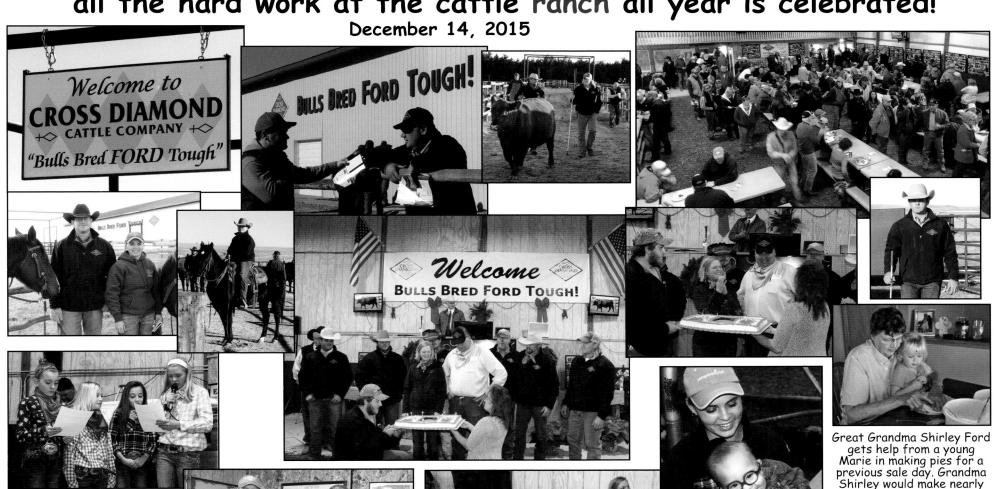

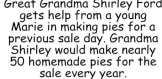

Rianna and Sheridan help Johanna and Marie Ford give the opening prayer before the sale!

Great Grandma Shirley Ford gets help from a young Marie in making pies for a previous sale day. Grandma Shirley would make nearly 50 homemade pies for the sale every year.

Cross Diamond Mission Statement: "We strive through respectful stewardship, to fulfill the responsibility bestowed on us to improve the land, cattle and people we serve, using God's blessings, to feed His people high quality beef."

Lee, Becky, Rianna and Sheridan Chaney relocated from Maryland to Nebraska in 2013. The Chaneys started writing their ag educational children's books in 2008 and are proud their books are in classrooms from coast to coast.

The Chaneys are passionate about agriculture and spreading their agricultural message. The girls love ranch life and are active in the community.

For more information about their ag presentations, or to order books, contact Becky at 308-785-8064, chaneyswalkabout@aol.com, info@rebeccalongchaney.com or check out her website at www.rebeccalongchaney.com

The Chaneys are proud their first seven children's books have received numerous "Agricultural Book of the Year" recognitions! The "Chaney Twins' Ag Series" and Lesson plans have been added to the USDA Agriculture in the Classroom Resource Directory as well as the American Farm Bureau's Ag Literacy List.

Lesson Plans are available free in downloadable form on the Pennsylvania Beef Council website at www.pabeef.org and at www.rebeccalongchaney.com
Box book discounts are available!

Kathy Moser Stowers, greenhouse plants and produce grower, lives with her husband, Tom and son, Wade on her family's second farm, Twin View Acres 2, near Jefferson, Maryland.

She has 25 years of experience in design and layout. She has shared her talents and expertise with local, state, national and international organizations and events.

For aid in designing brochures, pamphlets, program books, business cards, etc., contact Kathy at wakstowers@aol.com or call (301) 748-9112.

Award-winning photographer Kelly Hahn Johnson not only is known for her unique photojournalism style but also her approach to portraiture. She's won countless awards and her images have been featured in local, state and national publications. She loves observing people, moments and emotions, creating photos that will be treasured for years to come, like the priceless images she's captured of the Chaney twins over the years. She lives with her husband Blane and son Brady in a century-old renovated house in Sharpsburg, Maryland.

Visit her online gallery at kellyhahnphotography.com.
Contact Kelly at info@kellyhahnphotography.com
or call her at (240) 285-3677.

Rocky Mountain
Jewel of the Rockies

Written and photographed by
David Dahms

Bighorn ram in snowstorm

Front Cover: Hallett Peak and Bear Lake

Title Page: Summer sunrise

Back Cover: Longs Peak from Rock Cut
Foraging Pika
Mule deer fawn

Published by
Paragon Press
2360 Wapiti Road
Fort Collins, CO 80525

Text, photographs, and map
copyright © 1997 by David Dahms
All rights reserved.

No portion of this book may be reproduced
without written permission of the publisher.

ISBN 0-9646359-1-7

Printed in Singapore

    David Dahms is a nature and wildlife photographer, a career taken up after 16 years as a design engineer with Hewlett-Packard. His work has been published in a variety of magazines and calendars over the last ten years. His first book, *Rocky Mountain Wildlife*, was published in 1995.

    Dave uses a Canon A-1 35mm camera with lenses from 24mm to 600mm. Equipment specifics are insignificant compared to just being at the right place at the right time often enough until a spectacular situation occurs. Most of these times are near dawn or dusk, because of the warm, directional light and increased wildlife activity. There are also many days when nothing happens, but any day in the park is a good day regardless of the photographic outcome.

This color-relief map shows the major topological features of the park with its perspective view from the east. It was computer-generated from USGS elevation data.

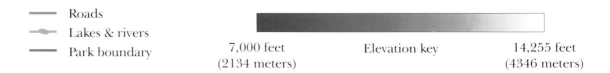

Dream Lake reflects a summer sunrise on Hallett Peak and Flattop Mountain. The 1.8 mile walk from Bear Lake to Nymph, Dream and Emerald Lakes is a very popular hike in the park.

Afternoon sunshine warms the jagged granite at Rock Cut and distant Longs Peak, while storm clouds build over the plains. At 14,255 feet above sea level, Longs Peak is the highest point in the park and a prominent landmark for many miles. Longs Peak was named for Major Stephen Long, the leader of an 1820 expedition in the area.

Old Man Winter is reluctant to release his icy grip on the high country, as demonstrated by this storm on Memorial Day. Fresh snow blankets the twisted trees near timberline, while fog and clouds obscure Mount Chapin across Fall River canyon.

A wet spring storm leaves a glorious frosting of pristine snow on the rocks and trees near Bear Lake and Hallett Peak. The distinctive shape and central location of Hallett Peak make it a major landmark of the park.

The appearance of the Pasqueflower heralds the coming of Spring. It is one of the first flowers to bloom early in the season. Its name is derived from "pâques," the French word for Easter.

As twilight fades into darkness, a sleepy great horned owl prepares for its nighttime hunting duties. Owls have very sensitive eyes and ears so they can find their prey in the dark. This owl is perched on the gnarled roots of a fallen ponderosa pine in Moraine Park.

Chasm Falls on Fall River surges to life with spring runoff as melting snow turns quiet creeks and streams into rushing torrents. Chasm Falls is about a mile up Fall River Road and can be reached by a spring hike or a summer drive.

Tiny pale aspen leaves sprout anew as springtime returns to the park, while Glacier Creek rushes with snowmelt.

Although the calendar says summer, the high mountains visible from Trail Ridge Road remain covered with winter's snow. Four peaks can be seen here: Mount Julian (12,928 ft), Cracktop (12,760 ft) with its namesake snowfilled crack, Chief Cheley Peak (12,804 ft), and Mount Ida (12,880 ft). Arrowhead Lake remains frozen and snow-covered in the basin near the bottom.

Morning sun lights Ypsilon Mountain over a layer of valley fog. Rising to 13,514 feet, it dominates the skyline northwest of Horseshoe Park. The mountain was named because of the Y-shaped crack in its face which resembles the Greek letter Ypsilon.

In the Lumpy Ridge area in the northeastern corner of the park is tiny Gem Lake. The lake is nestled in a rocky amphitheater with a sandy beach. The trail passes through alternating aspen and conifer groves and winds around whimsical rounded rock formations.

Poudre Lake lies beside the Continental Divide at Milner Pass and is the source of the Cache la Poudre River. The Continental Divide is a line across the backbone of the continent which separates streams that drain to the Atlantic and Pacific Oceans.

Spring is the time when baby animals are born in the park. This newborn elk calf struggles to stand up on its wobbly legs while its mother rests. There are approximately 3,000 elk in the park.

In mid-summer, placid Cub Lake is adorned with water lilies. Their floating leaves and large yellow blossoms decorate the shallow areas of the lake, while Stones Peak stands in the distance. Cub Lake is an easy and beautiful hike in the Moraine Park area, with many trailside flowers along the way.

Icy Brook tumbles from The Loch down this long, lovely cascade. In mid-summer, its banks are trimmed with yellow flowers of arrowleaf groundsel and an occasional white cow parsnip.

The red-shafted flicker is a common large woodpecker, named for the salmon color of its wing linings. A pair of flickers will excavate a cavity in a tree to build their nest. After the eggs hatch, both parents search for food for their nestlings. They are often seen on the ground, gobbling up ants. The chicks grow rapidly and get quite large before leaving the nest.

A morning rain shower soaks upper Fall River Canyon, while the sun greets us with a rainbow. The unpaved, narrow and winding Old Fall River Road runs up this valley to join Trail Ridge Road.

In the summertime, food is plentiful and the living is easy for the wildlife. This bull elk is dining on a fresh salad of lush streamside plants. In the summer the growing antlers of bull elk are nourished by a furry covering called velvet, which will dry up and fall off by autumn.

The dainty Blue Columbine is Colorado's state flower.

The Wood Lily is as beautiful as it is rare.

Wild Rose (also Woods Rose)

Showy Daisy (also Aspen Daisy)

Sego Lily (also Mariposa Lily)

Golden Banner (also Golden Pea)

The twisted trees at timberline lead a tortured existence. They only have a few short weeks of summer to grow. The rest of the year they must endure the brutal alpine winter. Branches can only grow on the downwind side of their trunks or sheltered behind rocks because raging winter storms scour their limbs with wind-driven ice pellets.

With a little patience, tiny pikas can be seen scampering among jagged tundra rocks. All summer they diligently collect mouthfuls of plants and carry them to storage spaces under the rocks. Pikas are related to rabbits and do not hibernate in winter. Instead they spend the time among the rocks, under the blanket of snow, eating their cache of dried plants.

Sunrise sets the sky ablaze over Sprague Lake, a picnic area midway along the Bear Lake road. This lake was built by Abner Sprague, who came to Colorado with his family in 1864 at age 15. Over the years he was an innkeeper, guide, explorer, surveyor, cartographer, and a prominent figure in the history of the park. He assigned names to many features of the area.

First light illuminates the east buttress of Taylor Peak and reflects in the calm waters of The Loch.
It is a lovely subalpine lake set in a high valley called Loch Vale.

The waters of Glacier Creek tumble over the round granite boulders of Alberta Falls, a short, easy, and popular hike from Glacier Gorge Junction. Alberta Falls was named by Abner Sprague for his wife, Alberta.

This male mountain bluebird pauses in the doorway of his nest in a ponderosa pine tree. Mountain bluebirds typically live in higher elevation areas above 8,000 feet. Each June, pairs of bluebirds build their nests in existing cavities in trees. The male waits patiently in a nearby tree while the female incubates the eggs. After the eggs hatch, both parents busily search for insects to feed their hungry chicks.

Mills Lake lies high in Glacier Gorge with a commanding view of Longs Peak, 4,300 feet above. The lake's name is a tribute to Enos Mills, a naturalist, guide, and early resident of the area, who was instrumental in the campaign to establish this park. These efforts were finally rewarded on January 26, 1915, when President Woodrow Wilson signed the bill officially creating Rocky Mountain National Park.

Lake Helene is close to treeline at the crest of Odessa Gorge. Windswept subalpine fir trees line its shores while Notchtop Mountain towers impressively overhead. The Odessa Lake trail from Bear Lake climbs steadily to this point before descending rapidly to Odessa Lake.

Odessa Lake lies in Odessa Gorge between Bear Lake and Moraine Park, flanked by dramatic mountain peaks. This rocky precipice called The Gable overlooks the lake on the north.

Fern Lake is less than a mile downstream from Odessa Lake, in a dense forest of lodgepole pine. Notchtop Mountain, Little Matterhorn, and The Gable dominate the skyline. There are wonderful wildflowers alongside the first several miles of the Fern Lake trail, including the namesake pteridium ferns.

The Clark's nutcracker and Steller's jay are two common birds in the park. The Steller's jay is deep blue, with a black head and crest. It is the high country relative of the eastern blue jay, and is similarly noisy and boisterous. The gray Clark's nutcracker feeds on the cones of the limber pine tree. It buries some seeds in underground caches, which helps propagate the trees. It got its name from William Clark of the Lewis and Clark expedition, who discovered it in 1805.

The golden-mantled ground squirrel and least chipmunk are two small mammals often seen scampering around in the park. Both spend the winter hibernating in underground burrows. The chipmunk can be discerned from the squirrel because the chipmunk is smaller and has stripes on its head.

Roaring River seems like an extreme name for this tiny stream, gently flowing over Horseshoe Falls. It is hard to believe it could have carved the huge channel which contains it now. On July 15, 1982, an old dam on Lawn Lake burst and 300 million gallons of water surged downhill. The torrent dislodged tons of debris, snapped trees, and created an alluvial fan which buried the road under 40 feet of rubble. The huge car-sized boulders scattered about are a testament to the power of rushing water.

Broad-tailed hummingbirds are nimble and precise aviators, hovering in place while sipping nectar from flowers with their long beaks. The female hummingbird builds her tiny nest of lichens and spider web, lays and incubates her eggs, and feeds her chicks, all with no help from the male. An adult hummingbird is about four inches long and weighs four grams, about as much as a single penny.

Marmots are large furry rodents often seen sunning themselves on prominent rocks, like this mother marmot with her baby. They can be found throughout the park, from low meadows to high alpine tundra. Throughout the summer they gorge themselves on plants and vegetation to fatten up for their long winter hibernation.

This tiny meandering stream in Kawuneeche Valley along the west side of the park is the headwaters of the mighty Colorado River. It will be joined by many other rivers and pass through the Grand Canyon on its 1,440 mile trip to the Gulf of California.

By mid-September the aspen leaves have turned to gold, the air is crisp and cool, and the hiking trails are even more enticing.

As summer comes to an end, streams that were rushing with snowmelt dwindle to a trickle, and the aspen leaves begin to lose their chlorophyll green.

Puffy evening clouds drift lazily over Lake Irene, a fishing and picnic area west of Milner Pass along Trail Ridge Road.

Stars seem especially plentiful and conspicuous in the night sky above the park. There are two reasons for this—there is very little stray light to mask the faint stars, and at high altitude there is less atmosphere to diffuse the star light. This photo is a seven-hour exposure, where the earth's rotation makes the stars appear to turn in the sky over the Mummy Range.

These trees are called "quaking aspen" because their leaves flutter in the slightest breeze. Even the scientific name, *populus tremuloides*, means trembling poplar. Elk eat aspen bark in the winter, leaving black scars on the trunks.

The shortened days of autumn cause aspen leaves to turn to beautiful fall colors in September, but not all trees change at the same time. This is because aspen propagate mainly with underground sucker roots instead of seeds, so the trees in each group are all genetically identical. However, an adjacent unrelated group may change at a different time.

Trail Ridge Road is a major attraction of the park. It is the highest continuous paved highway in the nation, reaching 12,183 feet above sea level in its 48-mile course from Estes Park to Grand Lake. Just west of Rainbow Curve it passes through this area of twisted and distorted trees at timberline before climbing to the treeless alpine tundra. The road is closed by snow for much of the year. It was built between 1929 and 1933 as an alternative to the narrow and winding Fall River Road.

This spectacular panorama of mountain peaks treats the hiker on the Bierstadt Lake trail. Thatchtop, Mount Otis, and Hallett Peak form the skyline, while the road to Bear Lake runs through the valley below. The trail switchbacks up the side of Bierstadt Moraine, winding through groves of twisted aspen, to reach Bierstadt Lake on top of the moraine. These features were named for 1870's artist Albert Bierstadt, who painted Longs Peak and other grand scenes of the American West.

The park meadows echo with the bellowing sounds of bugling elk in the fall. September is the rutting season for the elk, and it draws a large contingent of wildlife watchers. The largest bulls collect groups of cows and defend them from other bulls. They strut around, showing off their magnificent antlers and bugling to claim territories and attract cows. Usually a large bull will repel a smaller one with bluff and bluster, though occasionally two bulls will battle with locked antlers.

Golden aspen reflect in Fan Lake at sunrise. This lake was created by the 1982 Lawn Lake flood, when the deluge of flood waters washed tons of debris down into Horseshoe Park and dammed Fall River.

Late autumn is rutting season for the mule deer, and the bucks start paying a lot of attention to the does. Rival bucks usually determine dominance by intense threats and posturing, but sometimes engage antlers and hold a spectacular pushing and shoving battle. Mule deer are named for their large mule-like ears.

The Diamond on Longs Peak shines at a crisp autumn sunrise, while the pine trees remain dusted from an overnight snow shower. The 945-foot sheer upper east face of Longs Peak is nicknamed "the Diamond" and offers a formidable challenge to technical mountain climbers.

Boulder Brook begins high in the Boulder Field near Longs Peak. By autumn, it has slowed to a trickle and gently tumbles down this dainty cascade, lined with lush vegetation.

The majestic Rocky Mountain bighorn sheep is the state animal of Colorado and the official symbol of the park. They are well-known for their dramatic head-butting battles. Groups of females (ewes) stay separate from the males (rams) for most of the year. During the summer they may come to Sheep Lake in Horseshoe Park to lick salt and minerals. Both sexes have horns, but only rams develop massive curled horns. Rams' social hierarchy depends on horn size and a complex system of posturing and body language. When that fails, they resort to their familiar head-bashing fights.

Clouds billow over Specimen Mountain after an autumn snow. It is of volcanic origin and was named because of the mineral samples that were once collected there. The Cache la Poudre River meanders through the valley below.

Just west of Specimen Mountain is an area of rocky cliffs called The Crater. This point can be reached by a steep hike from Milner Pass, but you cannot go any further. The area is closed to hikers to protect the bighorn sheep that live on these slopes. The mountains of the Never Summer Range fill the skyline.

In mid-winter, frosty Hallett Peak and Flattop Mountain stand silently over frozen and snow-covered Nymph Lake. Although buried by many feet of snow, most of the hiking trails can be traveled on snowshoes or crosscountry skis.

This dramatic view rewards the winter hiker along the trail to Nymph Lake. The block summit of Longs Peak, the jagged spires of the Keyboard of the Winds, and Pagoda Mountain rise over frigid Glacier Gorge.

Morning sunshine briefly illuminates Flattop Mountain while a cloud deck threatens from the west. Although it may be flat on top, the southeast side of Flattop Mountain has rugged spires and cliffs.

The sunny east face of Taylor Peak shines behind the rounded pinnacles of The Sharkstooth.

A glorious winter sunset reflects on the frozen surface of Sprague Lake, silhouetting Mount Otis, Hallett Peak, and Flattop Mountain.